Contents

Preface

About the Author

The Plant Kingdom

Preface

Welcome to the Plant Kingdom. This book describes general characteristics of algae, bryophyte, pteridophyte, gymnosperms, and angiosperms.

I would like to thank family members and friends. The author invites suggestions from readers for the improvement of text.

The Plant Kingdom

The scientific study of plants are called botany. Plants included algae, bryophyte, pteridophyte, and gymnosperms. There are at least 320,000 species of plants are found. Humans depends on plants for food, medicine.

Algae:

i. They are holophytic organisms.

ii. They are found in fresh water, ponds, and pools.

iii. They do not contains true roots, stem and leaves.

iv. They are chlorophyll bearing plants.

v. The plant body may be unicellular or multicellular.

vi. Algae are capable of photosynthesis.

vii. They are heterotrophic.

Some examples of algae are given below-

- Nostoc.
- Chara.
- Marimo.
- Vaucheria.
- Diatom.
- Volvox.

Bryophyte:

i. Bryophytes are non vascular plants.

ii. Bryophytes are amphibious plants.

iii. Bryophytes are terrestrial plants but without water they do not complete their life cycle.

iv. Bryophytes come from two greek words "bryon" means moss and "phyton" means plants.

v. The plant body is thalloid i.e not differentiated into true roots, stems, and leaves.

vi. The plants are green.

vii. They may be autotrophic or heterotrophic.

viii. Xylem and phloem are completely absent.

ix. The male sex organs are called antheridia while the female sex organs are called archegonia.

x. The sporophytes are differentiated into foot, seta, and capsule.

Some examples of Bryophytes –

- Riccia.
- Marchantia.
- Pelia.
- Porella.
- Anthoceros.
- Sphagnum.

Pteridophyte:

i. Pteridophytes are come from two greek words pteron means feather and phyton means plants.

ii. There are about 11,000 species in pteridophytes.

iii. Pteridophytes are bounded by two layers i.eexine and initine.

iv. The plant body is differentiated into roots, stems and leaves.

v. The pteridophytes are vascular plants i.e xylem and phloem are present in pteridophytes.

vi. The life cycle of pteridophytes includes alternation of generations.

Some examples of pteridophytes includes-

- Psilopsida.
- Lycopsida.
- Pteropsida.

Gymnosperms:

i. Gymnosperms come from two Greek words "Gymnos" means naked and sperma means seeds.

ii. Plants are trees, shrubs, or lianes.

iii. Xylem and phloem are present in gymnosperms.

iv. Xylem is made up of tracheids, parenchyma, and rays while phloem is made up of sieve cells.

v. Plants have a tap root system.

vi. Ovule consist of a single intugement.

vii. Endosperms is haploid.

Some examples of gymnosperms are-

- Cycas.
- Pinus.
- Ginko.
- Gnetum.

Angiosperms:

i. Angiosperms come from the Greek words "Angeion'' means case and "sperma ,means seeds.

ii. There are at least 30,000 species in angiosperms.

iii. The flowering plants are known as angiosperms.

iv. Angiosperms are seed bearing plants.

v. The reproductive structures are called angiosperms.

vi. Angiosperms are trees, shrubs, or herbs.

vii. They are found in any places like forests, and grasslands.

viii. Angiosperms bears seeds, flowers, and fruits.

Some examples of angiosperms are-

- Magnolia Trees.
- Rose.

- Tulip.
- Tomatoes.

Classifications of Angiosperms:

Angiosperms are divided into two classes-

i. Monocotyledonous Plant.

ii. Dicotyledonous Plant.

Monocotyledonous Plant:

i. Presence of adventitious roots.

ii. The number of vascular bundles is more and closed.

iii. Flowers are trimerous.

Some examples of monocotyledonous plants are-

- Bamboos.
- Banana.
- Cereals.

Dicotyledonous Plant:

i. Two cotyledons are present.

ii. Flowers can be tetramerous or pentamerous.

Some examples of dicotyledonous plants are-

- Grapes.
- Sunflower.
- Tomato.
- Potato.

References:

https://www.jagranjosh.com/general-knowledge/classification-of-plant-kingdom-1453445359-

https://www.vedantu.com/biology/plant-kingdom-plantaehttps://en.wikipedia.org/wiki/Plant

https://en.m.wikipedia.org/wiki/

https://www.britannica.com/science/algae/Ecological-and-commercial-importance

https://www-livescience-com.cdn.ampproject.org/v/s/www.livescience.com/amp/54979-what-are-algae.html?usqp=mq331AQRKAGYAfj1idLKyZ6YwwGwASA%3D&amp_js_v=a2&amp_gsa=1#referrer=https%3A%2F%2Fwww.google.com&ampshare=https%3A%2F%2Fwww.livescience.com%2F54979-what-are-algae.html

Angiosperm Phylogeny Group (APG). 1998. An ordinal classification for the families of flowering plants. Annals of the Missouri Botanical Garden 85: 531-553.

Angiosperm Phylogeny Group (APG II). 2003. An update of the Angiosperm Phylogeny Group classification for the orders and families of flowering plants: APG II. Botanical Journal of the Linnean Society 141: 399-436.

Baillon, H. 1869. Histoires des plantes, vol. 1. L. Hachette & Cie. Paris, France.

Barkman, T. J., G. Chenery, J. R. Mcneal, J. Lyons-Weiler, and C. W. Depamphilis. 2000. Independent and combined analyses of sequences from all three genomic compartments converge on the root of flowering plant phylogeny. Proceedings of the National Academy of Sciences, USA 97: 13166-13171.

Barkman, T. J., S.-H. Lim, K. M. Salleh, and J., Nais. 2004. Mitochondrial DNA sequences reveal the photosynthetic relatives of Rafflesia, the world's largest flower. Proceedings of the National Academy of Sciences, USA 101: 787-792.

Barraclough, T. G. and V. Savolainen. 2001. Evolutionary rates and species diversity in flowering plants. Evolution 55:677-683.

Basinger, J., and D. L. Dilcher. 1984. Ancient bisexual flowers. Science 224: 511-513.

Behnke, H.-D. 1969. Die Siebrohren-PlastidenbeiMonocotlen. Naturwissenschaften 55: 140-141.

Beck, C. B., ed. 1976.Origin and Early Evolution of Angiosperms. Columbia University Press, New York.

Bell, C. D., D. E. Soltis, and P. S. Soltis. 2005. The age of the Angiosperms: a molecular timescale without a clock. Evolution, 59(6).

Dahlgren, R. 1980. A revised system of classification of the angiosperms. Botanical Journal of the Linnean Society 80: 91-124.

Dahlgren, R., H. Clifford, and P. Yeo. 1985. The families of the monocotyledons: structure, evolution and taxonomy. Springer-Verlag, Berlin, Germany.

Davies, T. J., T. G. Barraclough, M. W. Chase, P. S. Soltis, D. E. Soltis, And V. Savolainen. 2004. Darwin's abominable mystery: insights from a supertree of the angiosperms. Proceedings of the National Academy of Sciences, USA 101: 1904-1909.

Davis, C. C., and M. W. Chase. 2004. Elatinaceaeare sister to Malpighiaceae; Peridiscaceae belong to Saxifragales. American Journal of Botany 91: 262-273.

Davis, J. I., M. P. Simmons, D. W. Stevenson, and J. F. Wendel. 1998. Data decisiveness, data quality, and incongruence in phylogenetic analysis: an example from the monocotyledons using mitochondrial atpA sequences. Systematic Biology 47: 282-310.

Davis, J. I., D. W. Stevenson, G. Petersen, O. Seberg, L. M. Campbell, J. V. Freudenstein, D. H. Goldman, C. R. Hardy, F. A. Michelangeli, M. P. Simmons, C. D. Specht, F. Vergara-Silva, and M. A. Gandolfo. A phylogeny of the monocots, as inferred from rbcL and atpA sequence variation. Systematic Botany: in press.

De Jussieu, A. L. 1789. Genera plantarumsecundumordinesnaturalsdisposita. Heissant and Barrois, Paris, France.

Dilcher, D. L. 1989. The occurrence of fruits with affinities to Ceratophyllaceae in lower and mid-Cretaceous sediments. American Journal of Botany 76: 162.

Dilcher, D. L., and Crane, P. R. 1984. Archaeanthus: an early angiosperm from the Cenomanian of the western interior of North America. Annals of the Missouri Botanical Garden 71: 351-383.

Donoghue, M. J., and J. A. Doyle. 1989. Phylogenetic studies of seed plants and angiosperms based on morphological characters. In K. Bremer and H. Jörnvall [eds.], The hierarchy of life: molecules and morphology in phylogenetic studies, 181-193. Elsevier Science Publishers, Amsterdam, The Netherlands.

Douglas, A., and S. C. Tucker. 1996. Comparative floral ontogenies among Persoonioideae including Bellendena (Proteaceae). American Journal of Botany 83: 1528-1555.

Doyle, J. A. 1998. Phylogeny of vascular plants. Annual Review of Ecology and Systematics 29:567-599.

Doyle, J. A., and M. J. Donoghue. 1986. Seed plant phylogeny and the origin of the angiosperms: an experimental cladistic approach. Botanical Review 52: 321-431.

Doyle, J. A., and M. J. Donoghue. 1992. Fossils and seed plant phylogeny reanalzyed. Brittonia 44: 89-106.

Doyle, J. A. and M. J. Donoghue. 1993. Phylogenies and angiosperm diversification. Paleobiology 19:141-167.

Doyle, J. A., M. J. Donoghue and E. A. Zimmer. 1994. Integration of morphological and rRNA data on the origin of angiosperms. Annals of the Missouri Botanical Garden 81:419-450.

Doyle, J. A., and P. K. Endress. 2000. Morphological phylogenetic analyses of basal angiosperms: comparison and combination with molecular data. International Journal of Plant Sciences 161 (Supplement): S121-S153.

Doyle, J. A., H. Eklund, and P. S. Herendeen. 2003. Floral evolution in Chloranthaceae: implications of a morphological phylogenetic analysis. International Journal of Plant Sciences 164 (Supplement): S365-S382.

Drinnan, A. N., P. R. Crane, and S. B. Hoot. 1994. Patterns of floral evolution in the early diversification of non-magnoliiddicotyledons (eudicots). Plant Systematics and Evolution 8 (Supplement): 93-122.

Duvall, M. R., G. H. Learn, L. E. Eguiarte, and M. T. Clegg. 1993. Phylogenetic analysis of rbcL sequences identifies Acoruscalamus as the primal extant monocotyledon. Proceedings of the National Academy of Sciences, USA 90: 4611-4644.

Eichler, A. 1875-1878. Blüthendiagramme I/II.Engelmann, Leipzig, Germany.

Eklund, H., J. A. Doyle, and P. S. Herendeen. 2004. Morphological phylogenetic analysis of living and fossil Chloranthaceae. International Journal of Plant Sciences 165: 107-151.

Endress, P. K. 1986. Reproductive structures and phylogenetic significance of eztant primitive angiosperms. Plant Systematic Evolution 152: 1-28.

Endress, P. K. 2001. The flowers in extant basal angiosperms and inferences on ancestral flowers. International Journal of Plant Sciences 162: 1111-1140.

Endress, P. K., and A. Igersheim. 1997. Gynoecium diversity and systematics of the Laurales. Botanical Journal of the Linnean Society 125: 93-168.

Endress, P. K., and A. Igersheim. 2000a. The reproductive structures of the basal angiosperm Amborellatrichopoda (Amborellaceae). International Journal of Plant Sciences 161 (Supplement): S237-S248.

Endress, P. K., and A. Igersheim. 2000b. Gynoecium structure and evolution in basal angiosperms. International Journal of Plant Sciences 161 (Supplement): S211-S223.

Feild, T. S., M. A. Zweiniecki, T. Brodribb, T. Jaffre, M. J. Donoghue, and N. M. Holbrook. 2000. Structure and function of tracheary elements in Amborellatrichopoda. International Journal of Plant Sciences 161: 705-712.

Friis, E. M., K. R. Pedersen, and P. R. Crane. 2000. Reproductive structure and organization of basal angiosperms from the early Cretaceous (Barremian or Aptian) of western Portugal. International Journal of Plant Sciences 161 (Supplement): S169-S182.

Friis, E. M., K. R. Pedersen, and P. R. Crane. 2001. Fossil evidence of water lilies in the Early Cretaceous. Nature 410:357-360.

Friis, E. M., J. A. Doyle, P. K. Endress, and Q. Leng. 2003. Archaefructus–angiosperm precursor or specialized early angiosperm? Trends in Plant Science 8: 369-373.

Gandolfo, M. A., K. C. Nixon, and W. L. Crepet. 2002. Triuridaceae fossil flowers from the Upper Cretaceous of New Jersey. American Journal of Botany 89: 1940-1957.

Gandolfo, M. A., K. C. Nixon, and W. L. Crepet. 2004. Cretaceous flowers of Nymphaeaceae and implications for complex insect entrapment pollinations mechanisms in early Angiosperms. Proceedings of the National Academy of Sciences, USA: online.

Gaut, B. S., S. V. Muse, W. D. Clark, and M. T. Clegg. 1992. Relative rates of nucleotide substitution at the rbcL locus of monocotyledonous plants. Journal of Molecular Evolution 35: 292-303.

Givnish, T. J., and K. J. Sytsma. 1997. Consistency, characters, and the likelihood of correct phylogenetic inference. Molecular Phylogenetics and Evolution 7: 320-333.

Givnish, T. J., J. C. Pires, S. W. Graham, M. A. McPherson, L. M. Prince, T. B. Patterson, H. S. Rai, E. H. Roalson, T. M. Evans, W. J. Hahn, K. C. Millam, A. W Meerow, M. Molvray, P. J. Kores, H. E. O'Brien, J. C. Hall, W. J. Kress, and K. J. Sytsma. Phylogenetic relationships of monocots based on the highly informative plastid gene ndhF: evidence for widespread converted convergence. In In J. T. Columbus, E. A. Friar, C. W. Hamilton, J. M. Porter, L. M. Prince, and M. G. Simpson [eds.]. Monocots: Comparative biology and evolution. 2 vols. Rancho Santa Ana Botanic Garden, Claremont, California, USA, in press.

Goremykin, V., V. K. I. Hirsch-Ernst, S. Wölfl, and F. H. Hellwig. 2003. Analysis of the Amborellatrichopoda chloroplast genome sequence suggests that Amborella is not a basal angiosperm. Molecular Biology and Evolution 20: 1499-1505.

Gottsberger, G. 1977. Some aspects of beetle pollination in the evolution of flowering plants. Plant Systematics and Evolution 1: 211-226.

Gottsberger, G. 1988. The reproductive biology of primitive angiosperms. Taxon 37: 630-643.

Graham, S. W., and R. G. Olmstead. 2000. Utility of 17 chloroplast genes for inferring the phylogeny of the basal angiosperms. American Journal of Botany 87: 1712-1730.

Graham, S. W., P. A. Reeves, A. C. E. Burns, and R. G. Olmstead. 2000. Microstructural changes in noncoding chloroplast DNA: interpretation, evolution, and utility of indels and inversions in basal angiosperm phylogenetic inference. International Journal of Plant Sciences 161 (Supplement): S83-S96.

Heckman, D. S., D. M. Geiser, B. R. Eidell, R. L. Stauffer, N. L. Kardos, and S. B. Hedges. 2001. Molecular evidence for the early colonization of land by fungi and plants. Science 293: 1129-1133.

Qiu, Y.-L., J.-Y. Lee, F. Bernasconi-Quadroni, D. E. Soltis, P. S. Soltis, M. Zanis, E. Zimmer, Z. Chen, V. Savolainen, and M. Chase. 2000. Phylogeny of basal angiosperms: analyses of five genes from three genomes. International Journal of Plant Sciences 161 (Supplement): S3-S27.

Ray, J. 1703. Methodusplantarumemendataetaucta. Smith and Walford, London, UK.

Reichenbach, H. G. L. 1827-1829. Dr.Joh. Christ, Moessler'sHandbuch der Gewaechskunde, [ed.] vols. 2, 3. Hammerich, Altona, Germany.

Renner, S. S. 1999. Circumscription and phylogeny of the Laurales: evidence from molecular and morphological data. American Journal of Botany 86: 1301-1315.

Rodman, J. E., P. S. Soltis, D. E. Soltis, K. J. Sytsma, and K. G. Karol. 1998. Parallel evolution of glucosinolate biosynthesis inferred from congruent nuclear and plastid gene phylogenies. American Journal of Botany 85: 997-1006.

Rodrìguez-Trelles, F., R. Tarrìo, and F. J. Ayala. 2002. A methodological bias toward overestimation of molecular evolutionary time scales. Proceedings of the National Academy of Sciences, USA 99: 8112-8115.

Ronse De Craene, L. P., P. S. Soltis, and D. E. Soltis. 2003. Evolution of floral structures in basal angiosperms. International Journal of Plant Sciences 164 (Supplement): S329-S363.

Sanderson, M. J. and M. J. Donoghue. 1994. Shifts in diversification rate with the origin of angiosperms. Science 264:1590-1593.

Sandèrson, M., J., and J. A. Doyle. 2001. Sources of error and confidence intervals in estimating the age of angiosperms from rbcL and 18S rDNA data. American Journal of Botany 88: 1499-1516.

Sanderson, M. J. et al. 2004.

Sauquet, H., J. A. Doyle, T. Scharaschkin, T. Borsch, K. W. Hilu, L. W. Chatrou, and A. Le Thomas. 2003. Phylogenetic analysis of Magnoliales and Myristicaceae based on multiple data sets: implications for character evolution. Botanical Journal of the Linnean Society 142: 125-186.

Savolainen, V., M. W. Chase, C. M. Morton, D. E. Soltis, C. Bayer, M. F. Fay, A. De Bruijn, S. Sullivan, and Y.-L. Qiu. 2000. Phylogenetics of flowering plants based upon a combined analysis of plastid atpB and rbcL gene sequences. Systematic Biology 49: 306-362.

Savolainen, V., M. F. Fay, D. C. Albach, A. Backlund, M. Van Der Bank, K. M. Cameron, S. A. Johnson, M. D. Lledó, J.-C. Pintaud, M. Powell, M. C. Sheahan, D. E. Soltis, P. S. Soltis, P. Weston, W. M. Whitton, K. J. Wurdack, and M. W. Chase. 2000b. Phylogeny of the eudicots: a nearly complete familial analysis based on rbcL gene sequences. Kew Bulletin 55: 257-309.

Schneider, E. L. 1979. Pollination biology of the Nymphaeaceae.In D. M. Caron [ed.], Proceedings of the fourth international symposium on pollination, 419-430.Maryland Agricultural Experiment Station Special Miscellaneous Publication 1. College Park, Maryland, USA.

Soltis, D. E., and P. S. Soltis. 2004. Amborella NOT a "basal angiosperm"? Not so fast. American Journal of Botany: in press.

Soltis, D. E., P. S. Soltis, D. R. Morgan, S. M. Swensen, B. C. Mullin, J. M. Dowd, and P. G. Martin. 1995. Chloroplast gene sequence data suggest a single origin of the predisposition for symbiotic nitrogen fixation in angiosperms. Proceedings of the National Academy of Sciences, USA 92: 2647-2651.

Soltis, D. E., P. S. Soltis, D. L. Nickrent, L. A. Johnson, W. J. Hahn, S. B. Hoot, J. A. Sweere, R. K. Kuzoff, K. A. Kron, M. W. Chase, S. M. Swensen, E. A. Zimmer, S.-M. Chaw, L. J. Gillespie, W. J. Kress, and K. J. Sytsma. 1997. Angiosperm phylogeny inferred from 18S ribosomal DNA sequences. Annals of the Missouri Botanical Garden 84: 1-49.

Soltis, D. E., P. S. Soltis, M. E. Mort, M. W. Chase, V. Savolainen, S. B. Hoot, and C. M. Morton. 1998. Infering complex phylogeneis using parsimony: an empirical approach using three large DNA data sets for angiosperms. Systematic Biology 47: 32-42.

Soltis, D. E., P. S. Soltis, M. W. Chase, M. E. Mort, D. C. Albach, M. Zanis, V. Savolainen, W. J. Hahn, S. B. Hoot, M. F. Fay, M. Axtell, S. M. Swensen, L. M. Prince, W. J. Kress, K. C. Nixon, and J. S. Farris. 2000. Angiosperm phylogeny inferred from 18S rDNA, rbcL, and atpB sequences. Botanical Journal of the Linnean Society 133: 381-461.

Soltis, D. E., A. E. Senters, M. Zanis, S. Kim, J. D. Thompson, P. S. Soltis, L. P. Ronse De Craene, P. K. Endress, and J. S. Farris. 2003. Gunnerales are sister to other core eudicots: implications for the evolution of pentamery. American Journal of Botany 90: 461-470.

Soltis, D. E., P. S. Soltis, M. W. Chase, and P. K. Endress. 2004. Angiosperm phylogeny, classification, and evolution. Smithsonian Institution Press, Washington, DC, USA: in press.

Soltis, D. E., V. A. Albert, V. Savolainen, K. W. Hilu, Y-L. Qiu, M. W. Chase, J., S. Farris, S. Stefanovic, D. W. Rice, J. D. Palmer, and P. S. Soltis. Genome-scale data, angiosperm relationships, and "ending incongruence": a cautionary tale in phylogenetics. Trends in Plant Science, in press.

Soltis, P. S., D. E. Soltis, and M. W. Chase. 1999. Angiosperm phylogeny inferred from multiple genes as a tool for comparative biology. Nature 402:402-404.

Soltis, P. S., D. E. Soltis, M. J. Zanis, and S. Kim. 2000. Basal lineages of angiosperms: Relationships and implications for floral evolution. International Journal of Plant Science161(Supplement): S97-S107.

Soltis, P. S., D. E. soltis, V. savolainen, P. R. Crane, and T. G. Barraclough. 2002. Rate heterogeneity among lineages of tracheophytes: integration of molecular and fossil data and evidence for molecular living fossils. Proceedings of the National Acadamy of Sciences, USA 99: 4430-4435.

Soltis, P. S., D. E. Soltis, M. W. Chase, P. K. Endress, and P. R. Crane. 2004. The diversification of flowering plants. In J. Cracraft and M. J. Donoghue [eds.], The tree of life. Oxford University Press, Oxford, UK.

Stebbins, G. L. 1974. Flowering Plants: Evolution Above the Species Level. Harvard University Press, Cambridge, Massachusetts.Sun, G., D. L. Dilcher, S. Zheng, and Z. Zhou. 1998. In search of the first flower: A Jurassic angiosperm, Archaefructus, from northeast China. Science 282: 1692-1695.

Heywood, V. 1993.Flowering plants of the world.B.T. Batsford Ltd., London, UK.

Hickey, L. J., and A. D. Wolfe. 1975. The bases of angiosperm phylogeny: vegetative morphology. Annals of the Missouri Botanical Garden 62: 538-589.

Hillis, D. M. 1996. Inferring complex phylogenies. Nature 383: 130.

Hilu, K. W., T. Borsch, K. Muller, D. E. Soltis, P. S. Soltis, V. Savolainen, M. Chase, M. Powell, L. Alice, R. Evans, H. Sauquet, C. Neinhuis, T. Slotta, J. Rohwer, and L. Chatrou. 2003. Inference of angiosperm phylogeny based on matK sequence information. American Journal of Botany 90: 1758-1776.

Hoot, S. B., S. Magallón, and P. R. Crane. 1999. Phylogeny of basal eudicots based on three molecular data sets: atpB, rbcL, and 18S nuclear ribosomal DNA sequences. Annals of the Missouri Botanical Garden 86: 1-32.

Hufford, L. 1992. Rosidae and their relationships to other nonmagnoliiddicotyledons: A phylogenetic analysis using morphological and chemical data. Annals of the Missouri Botanical Garden 79: 218-248.

Igersheim, A. and P. K. Endress. 1998. Gynoecium diversity and systematics of the paleoherbs. Botanical Journal of the Linnean Society 127:289-370.

Judd, W. S., C. S. Campbell, E. A. Kellogg, P. F. Stevens, and M. J. Donoghue. 2002. Plant systematics: a phylogenetic approach. Sinauer Associates, Inc., Sunderland, Massachusetts, USA.

Kallersjo, M., J. S. Farris, M. W. Chase, B. Bremer, M. F. Fay, C. J. Humphries, G. Pedersen, O. Seberg, and K. Bremer. 1998. Simultaneous parsimony jackknife analysis of 2538 rbcL DNA sequences reveals support for major clades of green plants, land plants, seed plants and flowering plants. Plant Systematics and Evolution 213:259-287.

Kim, S., V. A. Albert, M.-J. Yoo, J. S. Farris, M. Zanis, P. S. Soltis, and D. E. Soltis.2004a. Pre-angiosperm duplication of floral genes and regulatory tinkering at the base of flowering plants. American Journal of Botany: in press.

Kim, S., D. E. Soltis, P. S. Soltis, M. J. Zanis, and Y. Suh. 2004b. Phylogenetic relationships among early-diverging eudicots based on four genes: were the eudicots ancestrally woody? Molecular Phylogenetics and Evolution 31: 16-30.

Kong, H.-Z., Z. Chen, and A.-M.Lu. 2002. Phylogeny of Chloranthus (Chloranthaceae) based on nuclear ribosomal ITS and plastid trnL-F sequence data. American Journal of Botany 89: 940-946.

Kramer, E. M., R. L. Dorit, and V. F. Irish. 1998. Molecular evolution of genes controlling petal and stamen development: Duplication and divergence within the APETALA3 and PISTILLATA MADS-box gene lineages. Genetics 149: 765-783.

Kuzoff, R. K., L. Hufford, and D. E. Soltis. 2001. Structural homology and developmental transformations associated with ovary diversification in Lithophragma (Saxifragaceae). American Journal of Botany 88: 196-205.

Les, D. H., E. L. Schneider, D. J. Padgett, M. Zanis, D. E. Soltis, And P. S. Soltis. 1999. Phylogeny, classification, and floral evolution of water lilies (Nymphaeales): a synthesis of non-molecular, rbcL, matK, and 18S rDNA data. Systematic Botany: 24: 28-46.

Lipok, B., A. A. Gardine, P. S. Williamson, and S. S. Renner. 2000. Pollination by flies, bees, and beetles of Nupharozarkana and N. advena (Nymphaeaceae). American Journal of Botany 87: 898-902.

Litt, A., and V. Irish. 2003. Duplication and diversification in the APETALA1/FRUITFULL floral homeotic gene lineage: implications for the evolution of floral development. Genetics 165: 821-833.

Lockhart, P. J. and D. Penny. 2005. The place of Amborella within the radiation of angiosperms. Trends Plant Sci 10:201–202.

Loconte, H., and D. W. Stevenson. 1991. Cladistics of the Magnoliidae. Cladistics 7: 267-296.

Mabberley, D. J. 1993. The plant book: a portable dictionary of the vascular plants. Cambridge University Press, Cambridge, UK.

Magallón, S., P. R. Crane, and P. S. Herendeen. 1999. Phylogenetic pattern, diversity, and diversification of eudicots. Annals of the Missouri Botanical Garden 86: 297-372.

Magallón, S., and M. J. Sanderson. 2001. Absolute diversification rates in angiosperm clades. Evolution 55: 1762-1780.

Martin, W., D. Lydiate, H. Brinkmann, G. Forkmann, H. Saedler and R. Cerff. 1993. Molecular phylogenies in angiosperm evolution. Molecular Biology and Evolution 10:140-162.

Mathews, S., and M. J. Donoghue. 1999. The root of angiosperm phylogeny inferred from duplicate phytochrome genes. Science 286: 947-949.

Mathews, S., and M. J. Donoghue. 2000. Basal angiosperm phylogeny inferred from duplicate phytochromes A and C.

International Journal of Plant Sciences 161 (Supplement): S41-S55.

Bharathan, G., and E. A. Zimmer. 1995. Early branching events in monocotyledons–partial 18S ribosomal DNA sequence analysis. In P. J. Rudall, P. J. Cribb, D. F. Cutler, and C. J. Humphries [eds.], Monocotyledons: systematics and evolution, 81-107. Royal Botanic Gardens, Kew, London, UK.

Borsch, T., K. W. Hilu, D. Quandt, V. Wilde, C. Neinhuis, and W. Barthlott. 2003. Non-coding plastid trnT-trnF sequences reveal a highly supported phylogeny of basal angiosperms. Journal of Evolutionary Biology 15: 558-567.

Braun, A. 1864.Uebersicht der naturlichen Systems nach der Anordnungderselben. In P. Ascherson [ed.], Flora der Provinz Brandenburg, der Altmark und des Herzogthums Magdeburg, vol. 1, 22-67. Hirschwald, Berlin, Germany.

Bremer, K. 2000. Early Cretaceous lineages of monocot flowering plants. Proceedings of the National Acadamy of Sciences, USA 97: 4707-4711.

Bremer, K., A. Backlund, B. Sennblad, U. Swenson, K. Andreasen, M. Hjertson, J. Lundberg, M. Backlund, and B. Bremer. 2001. A phylogenetic analysis of 100+ genera and 50+ families of euasterids based on morphological and molecular data with notes on possible higher level morphological synapomorphies. Plant Systematics and Evolution 229: 137-169.

Burleigh, G. G., and S. Mathews. 2004. Phylogenetic signal in nucleotide data from seed plants: implications for resolving the seed plant tree of life. American.Journal of Botany. 91: 1599-1613

Chase, M. W., and V. A. Albert. 1998. A perspective on the contribution of plastid rbcL DNA sequences to angiosperm phylogenetics. In D. E. Soltis, P. S. Soltis, and J. J. Doyle [eds.], Molecular systematics of plants, vol. 2: DNA sequencing, 488-507. Kluwer, Boston, Massachusetts, USA.

Chase, M. W., D. E. Soltis, R. G. Olmstead, D. Morgan, D. H. Les, B. D. Mishler, M. R. Duvall, R. A. Price, H. G. Hills, Y.-L.Qiu, K. A. Kron, J. H. Rettig, E. Conti, J. D. Palmer, J. R. Manhart, K. J. Sytsma, H. J. Michaels, W. J. Kress, K. G. Karol, W. D. Clark, M. Hedren, B. S. Gaut, R. K. Jansen, K.-J. Kim, C. F. Wimpee, J. F. Smith, G. R. Furnier, S. H. Strauss, Q.-Y. Xiang, G. M. Plunkett, P. S. Soltis, S. M. Swensen, S. E. Williams, P. A. Gadek, C. J. Quinn, L. E. Eguiarte, E. Golenberg, G. H. Learn, Jr., S. W. Graham, S. C. H. Barrett, S. Dayanandan, and V. A. Albert. 1993. Phylogenetics of seed plants: an analysis of nucleotide sequences from the plastid gene rbcL. Annals of the Missouri Botanical Garden 80: 528-580.

Chase, M. W., D. E. Soltis, P. S. Soltis, P. J. Rudall, M. F. Fay, W. H. Hahn, S. Sullivan, J. Joseph, T. Givnish, K. J. Sytsma, and J. C. Pires. 2000. Higher-level systematics of the monocotyledons: An assessment of current knowledge and a new classification. In K. L. Wilson and D. A. Morrison [eds.], Monocots systematics and evolution, 3-16.CSIRO, Melbourne, Australia.

Chase, M. W., M. F. Fay, D. S. Devey, O. Maurin, J. Davies, Y. Pillon, G. Petersen, O. Seberg, M. N. Tamura, C. B. Asmussen, K. Hilu, T. Borsch, J. I. Davis, D. W. Stevenson, J. C. Pires, T. J. Givnish, K. J. Sytsma, and S. W. Graham. Multi-gene analyses of monocot relationships: a summary. In In J. T. Columbus, E. A. Friar, C. W. Hamilton, J. M. Porter, L. M. Prince, and M. G. Simpson [eds.]. Monocots: Comparative biology and evolution. 2 vols. Rancho Santa Ana Botanic Garden, Claremont, California, USA, in press.

Couper, R. A. 1958. British Mesozoic microspores and pollen grains: a systematic and stratigraphic study. Palaeontographica.Abteilung B, Paläophytologie 103: 75-179.

Crane, P. R., M. J. Donoghue, J. A. Doyle and E. M. Friis. 1989. Angiosperm origins. Nature 342:131.

Crane, P. R., E. M. Friis, and K.R. Pedersen. 1995. The origin and early diversification of angiosperms. Nature 374: 27-33.

Crane, P. R. and S. Lidgard. 1989. Angiosperm diversification and paleolatitudinal gradients in Cretaceous floristic diversity. Science 246:675-678.

Crepet, W., and K. Nixon. 1998. Fossil Clusiaceae from the Late Cretaceous (Turonian) of New Jersey and implications regarding the history of bee pollination. American Journal of Botany 85: 1122-1133.

Cronquist, A. 1981. An integrated system of classification of flowering plants. Columbia University Press, New York, USA.

Cronquist, A. 1988. The evolution and classification of flowering plants, 2nd ed. New York Botanical Garden, Bronx, New York, USA.

Nandi, O. I., M. W. Chase, and P. K. Endress. 1998. A combined cladistic analysis of angiosperms using rbcL and nonmolecular data sets. Annals of the Missouri Botanical Garden 85: 137 - 212.

Nickerson, J., and G. Drouin. 2004. The sequence of the largest subunit of RNA polymerase II is a useful marker for inferring seed plant phylogeny. Molecular Phylogenetics and Evolution 31: 403-415.

Nickrent, D. L., R. J. Duff, A. Colwell, A. D. Wolfe, N. D. Young, K. E. Steiner, and C. W. Depamphilis. 1998. Molecular phylogenetic and evolutionary studies of parasitic plants.In D. E. Soltis, P. S. Soltis, and J. J. Doyle [eds.], Molecular systematics of plants, vol. 2, 211-241..Kluwer, Boston, Massachusetts, USA.

Nickrent, D. L., A. Blarer, Y.-L.Qiu, D. E. Soltis, P. S. Soltis, and M. Zanis. 2002. Molecular data place Hydnoraceae with Aristolochiaceae. American Journal of Botany 89: 1809-1817.

Nickrent, D. L. and D. E. Soltis. 1995. A comparison of angiosperm phylogenies from nuclear 18S rDNA and rbcL sequences. Annals of the Missouri Botanical Garden 82:208-234. Parkinson, C. L., K. L. Adams, and J. D. Palmer. 1999. Multigene analyses identify the three earliest lineages of extant flowering plants. Current Biology 9: 1485-1488.

Perkins, J. 1925. Ubersichtuber die Gattungen der Monimiaceae. Engelmann, Leipzig, Germany.

Posluszny, U., and P. B. Tomlinson. 2003. Aspects of inflorescence and floral development in the putative basal angiosperm Amborellatrichopoda (Amborellaceae). Canadian Journal of Botany 81: 28-39.

Pyankov, V. I., E. G. Artyusheva, G. E. Edwards, C. C. J. Black, and P. S. Soltis. 2001. Phylogenetic analysis of tribe Salsoleae (Chenopodiaceae) based on ribosomal ITS sequences: implications for the evolution of photosynthesis types. American Journal of Botany 88: 1189-1198.

Qiu, Y.-L., J. Lee, F. Bernasconi-Quadroni, D. E. Soltis, P. S. Soltis, M. Zanis, Z. Chen, V. Savolainen, and M. W. Chase. 1999. The earliest angiosperms: evidence from mitochondrial, plastid and nuclear genomes. Nature 402: 404-407.

Stefanovic, S. D. W. Rice, and J. D. Palmer. 2004. Long branch attraction, taxon sampling, and the earliest angiosperms: Amborella or monocots? BMC Evolutionary Biology 4:35.

Sun, G., Q. Ji, D. L. Dilcher, S. Zheng, K. C. Nixon, and X. Wang. 2002. Archaefructaceae, a new basal angiosperm family. Science 296: 899-904.

Swensen, S. M. 1996. The evolution of actinorhizal symbioses: evidence for multiple origins of the symbiotic association. American Journal of Botany 83: 1503-1512.

Sytsma, K. J., and D. A. Baum. 1996. Molecular phylogenies and the diversification of angiosperms. In D. W. Taylor and L. J. Hickey [eds.], Flowering plant origin, evolution, and phylogeny, 314-340.Chapman and Hall, New York, USA.

Takhtajan, A. 1980. Outline of the classification of flowering plants (Magnoliophyta). Botanical Review 46: 225-359.

Takhtajan, A. 1987.System of Magnoliophyta.Academy of Sciences, Leningrad, USSR.

Takhtajan, A. 1991.Evolutionary Trends in Flowering Plants.New York; Columbia Univ. Press.

Takhtajan, A. 1997.Diversity and classification of flowering plants. Columbia University Press, New York, USA.

Taylor, D. and L. Hickey. 1992. Phylogenetic evidence for the herbaceous origin of angiosperms. Plant Systematics and Evolution 180:137-156.

Thorne, R. F. 1974. A phylogenetic classification of the Annoniflorae. Aliso 8: 147-209.

Thorne, R. F. 1992. Classification and geography of the flowering plants. Botanical Review 58: 225-348.

Tillich, H.-J. 1995. Seedling and systematics in monocotyledons. In P. J. Rudall, P. J. Cribb, D. F. Cutler, and C. J. Humphries [eds.], Monocotyledons: systematics and evolution, 303-352. Royal Botanic Gardens, Kew, London, UK.

Tomlinson, P. B. 1995. Non-homology of vascular organisation in monocotyledons and dicotyledons. In P. J. Rudall, P. J. Cribb, D. F. Cutler, and C. J. Humphries [eds.], Monocotyledons: systematics and evolution, 589-622. Royal Botanic Gardens, Kew, London, UK.

vanTieghem, P. 1897. Sur les Buxaées. Annales des Sciences NaturellesBotanique. Série 8, 5: 289-338.

Walker, J. W., and A. G. Walker. 1984. Ultrastructure of Lower Cretaceous angiosperm pollen and the origin and early evolution of flowering plants. Annals of the Missouri Botanical Garden 71: 464-521.

Warming, E. 1879. Haandbog I den systematiskebotanik. P. G. PhilipsensForlag, Copenhagen, Denmark.

Wikström, N., V. Savolainen, and M. W. Chase. 2001. Evolution of the angiosperms: calibrating the family tree. Proceedings of the Royal Society of London, B 268: 2211-2220.

Williams, P. S., and E. L. Schneider. 1993. Nelumbonaceae. In K. Kubitzki, J. Rohwer, and V. Bittrich [eds.], The families and genera of vascular plants, 470-473. Springer, Berlin, Germany.

Wilson, T. K. 1966. The comparative morphology of the Canellaceae. IV. Floral morphology and conclusions. American Journal of Botany 53: 336-343.

Yoo, M. J., C. D. Bell, P. S. Soltis, and D. E. Soltis. In press. Divergence Times and Historical Biogeography of Nymphaeales. Systematic Botany.

Young, D. A. 1981. Are the angiosperms primitively vesselless? Systematic Biology 6: 313-330.

Zanis, M., D. E. Soltis, P. S. Soltis, S. Mathews, and M. J. Donoghue. 2002. The root of the angiosperms revisited. Proceedings of the National Academy of Sciences, USA 99: 6848-6853.

Zanis, M. J., P. S. Soltis, Y.-L.Qiu, E. Zimmer, and D. E. Soltis. 2003. Phylogenetic analyses and perianth evolution in basal angiosperms. Annals of the Missouri Botanical Garden 90: 129-150.

Zhang, L.-B., and S. Renner. 2003. The deepest splits in Chloranthaceae as resolved by chloroplast sequences. International Journal of Plant Sciences 164 (Supplement) S383-S392.

Zimmer, E. A., R. K. Hamby, M. L. Arnold, D. A. LeBlanc and E. C. Theriot. 1989. Ribosomal RNA phylogenies and flowering plant evolution. Pp. 205-214 in The Hierarchy of Life, ed. B. Fernholm, K. Bremer and H. Jornvall. Amsterdam; Elsevier.

Zwickl, D. J., and D. M. Hillis. 2002. Increased taxon sampling greatly reduces phylogenetic error. Systematic Biology 51: 588-598.

www.ingramcontent.com/pod-product-compliance
Lightning Source LLC
Chambersburg PA
CBHW040340220526
45473CB00009B/2745